COLLECTED WRITINGS
OF
EDWARD LEEDSKALNIN

COLLECTED WRITINGS
OF
EDWARD LEEDSKALNIN

MAGNETIC CURRENT

&

A BOOK IN EVERY HOME

MOCKINGBIRD
—— PRESS ——

Cover, Copyright © 2023 Mockingbird Press LLC
Foreword by Elizabeth Ledbetter

Publisher's Cataloging-In-Publication Data

Leedskalnin, Edward, author; with Ledbetter, Elizabeth, foreword by
Collected Writings of Edward Leedskalnin: Magnetic Current & A Book in Every Home / Edward Leedskalnin; with Elizabeth Ledbetter

Paperback	ISBN-13: 978-1-68493-179-8
Hardback	ISBN-13: 978-1-68493-180-4
Ebook	ISBN-13: 978-1-68493-181-1

1. Science—Physics—Magnetism. 2. Philosophy—Essays. 3. Political Science—Commentary & Opinion. 4. Society & Social Sciences—Society & Culture: General—Popular Beliefs & Controversial Knowledge. I. Edward Leedskalnin. II. Elizabeth Ledbetter. III. Collected Writings of Edward Leedskalnin. IV. Title: Magnetic Current & A Book in Every Home.

SCI038000 / PHI035000 / POL046000 / JBG

Type Set in Century Schoolbook / **Franklin Gothic Demi**

Mockingbird Press, Augusta, GA
info@mockingbirdpress.com

Contents

FOREWORD

THE Collected Writings of Edward Leedskalnin is a compilation of two of Leedskalnin's works, Magnetic Current and A Book in Every Home. This eccentric sculptor and amateur scientist devoted most of his life to creating a large complex of megalithic stones that he quarried and carved himself.

Edward Leedskalnin was born in Latvia in 1887. Although he only received formal education up to the fourth grade, he was very inquisitive and spent a large part of his youth reading. He was said to be a sickly boy and grew into a small man—reportedly measuring just 5 feet tall and weighing 100 lbs.

At age 26, he was engaged to marry a 16-year-old girl named Agnes Skuvst, but the wedding was called off. Accounts differ, some claiming the wedding was canceled the day before it was scheduled and others saying that Leedskalnin was jilted at the altar. Regardless, he was heartbroken and shortly after emigrated to America.

After reaching New York in 1912, he continued on to Oregon where he worked for an ax-handle manufacturer. By 1923, he had contracted tuberculosis. The illness prompted him to move to a warmer climate for his health. He chose Florida, where he purchased an undeveloped acre of land in Florida City.

It was on this modest plot that he began Rock Gate (later renamed Coral Castle). This ambitious project involved extracting enormous pieces of oolite stone from his land, moving them into position, and carving them—entirely alone. The pieces include sculptures and carved stone furniture, as well as a two-story tower that served as his living quarters.

The project was devoted to his "Sweet Sixteen," the woman who had rejected him many years earlier.

Due to the scale of the project, some have dubbed it "Florida's Stonehenge." While it's unclear how Leedskalnin was able to complete it, it is all the more impressive considering his small stature and questionable health.

When he wasn't hewing or carving stone, Leedskalnin was also conducting experiments and writing. His first published work was a pamphlet titled A Book in Every Home, published in 1936. The book is divided into three parts, with the first section focusing on Leedskalnin's views on relationships and education. In the second section, Leedskalnin offers his opinions on domestic arrangements and the raising of children. And in the third, he shares his views on voting and the role of government, advocating that "the weaklings" should not be allowed to vote.

Leedskalnin also spent two years testing magnets and recording his findings from Rock Gate. These experiments would form the basis of Magnetic Current, published in 1945. In the book, Leedskalnin argues that electricity and magnetism are not separate phenomena but are instead two aspects of the same fundamental force, which he calls "magnetic current." He also proposes a new model of atomic structure and suggests that the fundamental particles of matter are tiny magnets that are constantly in motion.

Leedskalnin presents a variety of experiments throughout the book, many of which involve the manipulation of magnetic fields using simple items like various magnets, car batteries, light bulbs, and coils of wire. He believed that his discoveries could provide insights into the mysteries of the universe, including the relative positions of the celestial bodies.

Both works demonstrate Leedskalnin's unique view of the world and his eccentric personality. While some of his ideas may be seen as outdated or controversial, the books offer an intriguing glimpse into the mind of a self-taught philosopher and inventor who had a unique perspective on the world around him.

Magnetic Current

by

Edward Leedskalnin

Magnetic Current

THIS writing is lined up so when you read it you look East, and all the description you will read about magnetic current, it will be just as good for your electricity.

Following is the result of my two years' experiment with magnets at Rock Gate, seventeen miles Southwest from Miami, Florida. Between Twenty-fifth and Twenty-sixth Latitude and Eightieth and Eighty-first Longitude West.

First, I will describe what a magnet is. You have seen straight bar magnets, U shape magnets, sphere or ball magnets and Alnico magnets in many shapes, and usually a hole in the middle. In all magnets one end of the metal is North Pole and the other South Pole, and those which have no end one side is North Pole and the other South Pole.

Now about the sphere magnet. If you have a strong magnet you can change the poles in the sphere in any side you want or take the poles out so the sphere will not be a magnet any more. From this you can see that the magnet can be shifted and concentrated and also you can see that the metal is not the real magnet. The real magnet is the substance that is circulating in the metal. Each particle in the substance is an individual magnet by itself, and both North and South Pole individual magnets. They are so small that they can pass through everything. In fact they can pass through metal easier than through the air. They are in constant motion, they are running one kind of magnets against the other kind, and if guided in the right channels they possess perpetual power. The North and South Pole magnets they are cosmic force, they hold together this earth and everything on it. Each North and South Pole magnet is equal in strength, but the strength of each individual magnet doesn't amount to anything. To be of practical use they will have to be in great numbers.

In permanent magnets they are circulating in the metal in great numbers, and they circulate in the following way. Each kind of the magnets are coming out of their own end of the pole and are running around, and are running in the other end of the pole and back to its own end, and then over and over again. All the individual magnets do not run around. Some run away and never come back, but new ones take their place.

The earth itself is a great big magnet. In general, these North and South Pole individual magnets are circulating in the same way as in the permanent magnet metal. The North Pole individual magnets are coming out of the earth's South Pole and are running around in the earth's North Pole and back to its own pole, and South Pole individual magnets are coming out of the earth's North Pole and are running around, and in earth South Pole and back to its own end. Then both North and South Pole individual magnets start to run over and over again.

In a permanent magnet bar between the poles there is a semi-neutral part where there is not much going in or out, but on the earth there is no place where the magnets are not going in or out, but the magnets are running in and out at pole ends more than at the Equator. Now you get the equipment and I will tell you so you can see for yourself that it is in the way I have told. Get a permanent magnet bar four inches long. A U shape magnet that is strong enough to lift from ten to twenty pounds. An Alnico magnet about three inches long, two and one-half inches wide, one inch thick. Hole in the middle and poles in each end, several feet in length of hard steel fishing line. Line when it is not in coil it stays straight and a soft steel welding rod one-eighth of an inch thick and three feet long. From the fishing wire and the welding rod you will make magnets or compasses, and if you hang them up in fine threads by middle and keep them there they will be permanent magnets.

When you are making a magnet pole in the welding rod use U shape magnet. South Pole magnet to make North Pole magnet in the rod and use U shape North Pole magnet to make South Pole magnet in the rod. You can drag the magnet over the rod from end to end, but never stop in middle. If you stop in middle there will be an extra pole so it will disturb the magnet's circulation. Use iron filings to test the rod if there is any magnets in the middle, and if there is the filings will cling to it. Then drag the permanent magnet over the rod and it will take it out. To take the magnet out from rod ends approach or touch the rod end with the same kind of magnet that is in the rod, by dipping the rod ends in iron filings, you will see how it works.

Break three pieces of the steel fishing line just long enough to go in between the two poles of U shape permanent magnet. Put them endwise between the two poles, and take them out. Hang one by middle with fine thread, and hang it up in East side of the room where there is no other magnet or metal around. Now you will have a permanent magnet or compass to test the polarity in other magnets. For more delicate use hang the magnet in spider web. To test the strength of a magnet use iron filings.

Put the U shape permanent magnet two feet West from the hanging magnet. Hold the North Pole magnet in level with the hanging magnet, then you will see that the South pole of the hanging magnet is turning to you and the North Pole magnet away from you. Now put the South Pole permanent magnet pole in the same level, this time North Pole magnet will turn to you and South Pole magnet away from you. This experiment shows two things, one that the magnets can be sent out in straight streams, and the other whatever kind of magnets you are sending out the other kind of magnets are coming back to you.

Take two pieces of steel fishing line wire, put them in U shape magnet, hold a little while, take them out, bend a little back in one end and hang them up, and make it so that one magnet's lower end is North Pole magnet and the other South Pole magnet. Make it so that they hang three inches apart. Put North Pole North side, and South Pole South side. Now take the four-inch-long permanent magnet bar, hold North Pole in North side and South Pole in South side. Raise slowly up to the two hanging magnets, then you will see that the hanging magnets are closing up. Now reverse, put North Pole of bar magnet South side and South Pole North side. This time when bar magnet approaches the hanging magnets will spread out. This experiment shows that North and South Pole magnets are equal in strength and that the streams of individual magnets are running one kind of magnets against the other kind.

Cut a strip of a tin can about two inches wide and a foot long. Put the North Pole of the U shape magnet on top of the strip, and dip the lower end in iron filings, and see how much it lifts. Now put the South Pole on top and see how much it lifts. Change several times, then you will see that the North Pole lifts more than the South Pole. Now put the North Pole magnet under the iron filing box, and see how much it pushes up. Now change, put South Pole magnet under the box and see how much it pushes up. Do this several times, then you will see that the South Pole magnet pushes up more than North Pole magnet. This experiment shows again that on level ground the magnets are in equal strength.

Now take the three-foot long soft steel welding rod. It is already magnetized as a permanent magnet, hang it in a fine thread so it is in level. Now measure each and you will see that the South end is longer. In my location at Rock Gate, between Twenty-fifth and Twenty-sixth

Latitude and Eightieth and Eighty-first Longitude West, in three-foot long magnet the South Pole end is about a sixteenth of an inch longer. Farther North it should be longer yet, but at Equator both ends of the magnet should be equal in length. In earth's South hemisphere the North Pole end of magnet should be longer.

All my hanging magnets or compasses they never point to the earth's magnetic pole, neither to the geographical pole. They point a little Northeast. The only reason I can figure out why they point in that way is, looking from the same geographical meridian the North magnetic pole is on, the South magnetic pole is one hundred and fifteen longitudes West from it. In rough estimation the earth's South magnetic pole is two hundred and sixty miles West from the same meridian the earth's North magnetic pole is on. That causes the North and South Pole magnets to run in Northeast and Southwest direction. My location is too far away from the magnetic poles so all my magnets are guided by the general stream of individual North and South Pole magnets that are passing by.

Now I will tell you what magnetic current is. Magnetic current is the same as electric current is a wrong expression. Really it is not one current, they are two currents, one current is composed of North Pole individual magnets in concentrated streams and the other is composed of South Pole individual magnets in concentrated streams, and they are running one stream against the other stream in whirling, screwlike fashion, and with high speed. One current alone if it be North Pole magnet current or South Pole magnet current it cannot run alone. To run one current will have to run against the other.

Now I will tell you how the currents are running when they come out of a car battery, and what they can do. Now get the equipment. First put a wooden box on floor, open side up, cut two notches in middle so you can put a one-eighth of an inch thick and eighteen-inch long copper wire across the box. Put the wire one end East, the other West. Stay yourself West, put car battery South side of the box positive terminal East, negative terminal West, get two flexible leads and four clips to fit the battery and the bare copper wire, connect the East end of the copper wire with positive terminal, clip the West end of the copper wire with the West side flexible lead, leave the connection with negative terminal open.

Break two pieces of the steel fishing line one inch long, put each piece by middle across the copper wire, one on top of the copper wire and the other under, hold with your fingers, now touch the negative terminal with the loose clip, hold until the copper wire gets hot. Take them off, now you have two magnets, hang them up by middle in fine thread. The upper magnet will hang the way it is now, but the one below will turn around. Break five inches long piece of the fishing line, put the middle of the wire across and on top of the copper wire, touch the battery, hold until the copper wire gets hot, dip the middle of the wire in iron filings, then you will see how long a magnet can be made with this equipment.

Break or cut several pieces of the hard steel fishing wire as long as to go between the poles of the U shape magnet, now hold two pieces of the steel wire ends up and down, one wire South side of the copper wire, and the other North side, the lower ends just below the copper wire. Hold tight and touch the battery, hold until the copper wire gets hot, now hang them up by upper end just above the copper wire, touch battery, the South side magnet will swing South, and the North side magnet will swing North. Put two pieces on top of the copper wire, the ends just a little over the copper wire. Those ends lying on copper wire, one pointing South and the other North, hold tight, touch battery, hold until the copper wire gets hot, take off the one pointing South is South Pole magnet and the one pointing North is North pole magnet. Put one wire on top of the copper wire pointing South, other below pointing North. Magnetize, hang up by tail ends on the copper wire, touch battery they both will swing South. Put one wire on top of the copper wire pointing North, the other below pointing South, magnetize, hang up by tail end above the copper wire, touch the battery, both magnets will swing North.

Cut six pieces of fishing wire one inch long, put them by middle on top and across the copper wire. Hold tight, touch battery, hold until copper wire gets hot. Take off, now put glass over the copper wire, put those six pieces of magnets on glass, on top of the copper wire lengthwise just so the ends don't touch each other, touch the battery, they all will turn across the copper wire, now pull three to South side and three to North side in the same way, they lie now but about one-half of an inch away from the copper wire, touch battery, they all will jump on the copper wire. Now roll all six together, let loose, and you will see that they won't stay together. Magnetize one piece in U shape magnet, put North Pole end East on the copper wire, and South Pole West, touch the battery, the magnet will swing left. Now put South Pole East side and North Pole West side, this time the magnet will turn right, take glass off.

Take one piece of hard steel fishing wire, dip in iron filings and see there is no magnet in it. This time hold the wire up and down, the lower end on middle of the copper wire, hold tight. Touch the battery, hold until the copper wire gets hot. Take it off. Dip the wire in iron filings and you will see that it is no magnet. Why? To make magnets with currents from batteries and dynamos with a single wire the metal will have to be put on the wire in such a way so that the magnets which are coming out of the wire will be running in the metal starting from the middle of the metal and run to the end and not from end to middle and across as they did this last time. You have read that to make a South Pole in a coil end that is pointing to you, you will have to run positive electricity in the coil in clockwise direction. I can tell you that the positive electricity has nothing to do with making a South magnet pole in the coil. Each pole South or North is made by their own magnets in the way they are running in the wire. This magnet-making with a single wire, it illustrates how all magnets are made.

In a car battery the North Pole magnets run out of positive terminal and South Pole magnets run out of negative terminal. Both kinds of magnets are running, one kind of magnets against the other kind, and are running in the same right-hand screw fashion. By using the same whirling motion and running one kind of magnets against the other kind, they throw their own magnets from the wire in opposite directions. That is why if you put a magnet metal across the copper wire the one end is North Pole and the other end South Pole.

Get four pieces of wire size sixteen, six inches long, two copper and two soft iron, bend one end of each wire back so the clips can hold it better. Use copper wire first. Put both wires in clips, connect with battery, have the wire ends square, now put the loose ends together, and pull them away. Then you will notice that something is holding you back. What is it? They are magnets. When you put the ends together, the North and South Pole magnets are passing from one wire to the other, and in doing it they pull the wire ends together. Now put the soft iron wire in the clips, put the loose ends together, and pull them away. This time the passing magnets hold the wire ends together stronger. Put the ends together many times, then you will see which wire end gets red first, and which will make the bigger bubble in the end, and watch the little sparks coming out from the bubbles. Stretch the bubbles out while they are in liquid form, then you will see in the bubble that something is whirling around. Those little sparks you see coming out of the bubble, they are not the magnets, but the magnets are the ones which throw the sparks out of the bubbles. When all the magnets that are in the wire, if they cannot pass over to the other wire, they are expending the bubble and running out of it and carrying the metal sparks with them. When the bubble is cool, break it up, then you will see the space left where the magnets were in.

Get two pieces of lumber, one by six inches, a foot long, nail them together so that one lies flat on floor and the other on top the edges up and down. Cut a notch in end in upper piece, four inches deep and as high as to hold a piece of wood or brass that would hold needle points in ends and have a hole in middle to hold the three-foot magnet. Balance the magnet good so it would stop on its right magnetic position. Now put the car battery South side positive terminal East and negative terminal West. Connect the East end of the copper wire with positive terminal and connect the West end of the copper wire with the West side lead, hold the copper wire just above the magnet a quarter of an inch North of magnet's end, hold in level and square. Touch the battery, then you will see the magnet swinging East. Now put the battery North side, positive terminal East, negative terminal West, connect West end of the copper wire with negative terminal, connect East end of copper wire with East side lead, put the copper wire on top of the magnet a quarter of an inch South of magnet's end, hold the copper wire just above in square and level, touch the positive terminal, then you will see the magnet swinging West. If the battery is right, magnet strong enough, and the magnet rod balanced good it will repeat the same thing every time.

I think the batteries are not made right. Sometimes there is more of North Pole magnets than there is South Pole magnets. They should be equal, the same as from generators which do not run the South Pole magnets in frame or base, but run directly away the same as they run the North Pole magnets.

From the following experiment you will see that the battery is not balanced right. Put the copper wire across the box, one end East, the other end West, connect one lead a foot West from East end and the other lead with West end, hang a magnet in spider web, put the magnet in same level with the copper wire. Keep the copper wire end a little away from magnet's North Pole, connect East lead with positive terminal, tap the negative terminal several times with the loose clip, and see what the magnet is doing. Change the terminal, change the tapping, move the box and copper wire to the South Pole end, repeat the same thing. Then you will notice sometimes the copper wire end pushes away the North Pole magnet, and sometimes it pulls it in and the same thing happens with South Pole magnet, and sometimes it does nothing. So it shows the battery is irregular.

Connect the leads with battery's terminals to make a loop, keep the leads on the same level with battery, drag a hanging magnet over the loop and the connections between the battery's terminals. You will see that one end of the magnet keeps inside the loop, and the other outside, and the same thing happens when the magnet crosses the connection between the terminals. This experiment indicates that the North and South Pole magnet currents are not only running from one terminal to the other, but are running around in an orbit and are not only running one time around, but are running many times around until the North and South Pole individual magnets get thrown out of the wire by centrifugal force, and by crowding. While the North and South Pole magnets were in their own terminals they only possessed pushing power, the pulling power they acquire only if the other kind of magnets are in front of them, like the permanent magnets if you put the opposite magnet in front of it, then they will hold together. The same way you have done with the six inches long pieces of copper and soft iron wire.

From the experiment with the car battery you can see the principle how permanent magnets are made by North and South Pole individual magnet currents running in a single wire from battery. How did the magnets get in there? As I said in the beginning, the North and South Pole magnets they are the cosmic force, they hold together this earth and everything on it. Some metals and non-metals have more of the magnets than others. The North and South Pole magnets have the power to build up and take down, for instance in welding the magnets take the welding rod down and put it on the welding, in electro-plating they put one metal on the other, and if you burn a metal too much in an electric furnace the metal will disappear in air.

The North and South Pole magnets were put in the car battery by a generator. When the North and South Pole magnets went in the battery they built up a matter that held the magnets themselves, and later on the acid takes the matter in parts and separates the magnets and sends them to their own terminals, and from there they come out. In other batteries the acid takes the zinc in parts and sends the North Pole magnets to positive terminal and holds the South Pole magnets by itself for negative terminal. When the connections are made the magnets will come out of the battery and will come out until the zinc will last. When the zinc is gone the magnets are gone, too. The same is true if you put iron in acid and some other metals, for the other terminal and when the connections are made the magnets will come out of the battery, but when the iron is gone the magnets are gone, too. This should be sufficient to see that the North and South Pole magnets are holding together everything. You saw how magnetic currents are made in battery from metal by acid. Next I will tell you how magnetic currents are made by permanent and electric magnets, and then without either.

This time you will make an equipment that can be used for four purposes. Electric magnet, transformer, generator and holder of perpetual motion. Bend iron or soft steel bar one and one-half inch in diameter, bend in a U shape each prong a foot long, and three inches between the prongs, make two spools from brass or aluminum six inches long and big enough for the bar to go in. Wind fifteen hundred turns of insulated copper wire, size sixteen, on each spool. Put on as close to the bend as it will go. Connect the battery with the coils so that each current is running in both coils at the same time, and so that one end of the bar is North Pole and the other South Pole. Now you have an electric magnet.

This time the same thing will be a transformer. It will not be economical, it is only to show how a transformer works. Wind a coil of fifteen hundred turns with insulated copper wire, size eighteen, on a spool less than three inches long, so that one inch and a half square iron rod can go in easy, get two rods, one three, the other six inches long. If possible have them from laminated iron. Get two radio blue bead, six to eight-volt light bulbs. Now connect one light bulb with the three-inch coil, put the coil without a core between the loose ends of the iron prongs, connect the six-inch coils with battery, leave negative terminal open. Tap the negative terminal, then you will see the wire inside the light bulb turn red. Put iron core in the coil's hole, tap the battery, this time it will make light. Why did it not make just as much light the first time? The battery put just as much magnet in those iron prongs the first time as it did the last time, but as you see the coil did not get the magnets. Now you see the soft iron has a lot to do to make magnetic currents.

Magnetic currents, or if you want to call it electric current, make no light. We only get light if we put obstructions in the light bulbs. In the light bulbs the wire is so small that all magnets cannot pass through easily, so they heat the wire up and burn and make light. If the wire in the light bulb had been as large inside as it is outside then there would be no light. Then those individual magnets which are in the coil would dissipate in air.

Both North and South Pole individual magnet currents which came out of the car battery and went in the transformer were direct currents, but the light in the bulb was caused by alternating currents. (Have in mind that always there are two currents, one current alone cannot run. To run they have to run one against the other.) You transformed currents in kind. Now I will tell you how to transform currents in strength. To make higher voltage you wind the coil with smaller wire and more turns and to have less voltage wind the coil with bigger wire and less turns. The difference now is that this transformer makes alternating currents from direct currents and the power line transformers use alternating currents to make alternating currents in this transformer, the iron prong ends remain the same magnet pole, but in power line transformers the magnet poles alternate. In power line transformers the currents only are in motion and in this transformer the currents are in motion and you are, too.

Now about the generator. In the first place all currents are alternating. To get direct currents we have to use a commutator. Transformers and generators of any description are making the currents in the same way by filling the coil's iron core with magnets and letting the iron core push them out and into the coil. Connect the battery with the electric magnet, it will be a field magnet now. Put the three-inch coil between the iron prongs, and take it out, do it fast, repeat it, then you will have a steady light in the light bulb. Now you and the field magnet are a generator. Suppose you had a wheel and many coils around the wheel turning, then you would be making all kinds of light. Do not make the machine, I already have the application for patent in the Patent Office. I made ten different machines to make magnetic currents, but I found this combination between field magnets and coils the most efficient. Put the coil in slowly and take it out slowly, then you will have no light. That will show, to make magnetic currents, the time is important.

Put the six-inch long square rod on top of the two iron prongs, fit good so it lies even. Connect the battery with electric magnet for a little while, now disconnect the battery, connect the light bulb with the electric magnet the same way it was connected with the battery, now pull off the six-inch long bar, do it quickly, then you will see light in the bulb, connect the battery up again with the electric magnet, put the bar across the iron prongs, hold awhile, disconnect the battery. Now the electric magnet holds perpetual motion. If not disturbed it will last indefinitely. I held it in this position for six months, and when I pulled off the six-inch bar I got just as much light out of it as I got in the first time. This experiment shows that if you start the North and South Pole individual magnets in an orbit, then they will never stop. The hanging magnets that hang up and down, they show that there is motion inside the bar. Hold the perpetual motion holder North Pole magnet or pole end East and South Pole magnet terminal or pole end West, now raise it up slowly to the South Pole hanging magnet, then you will see the South Pole hanging magnet swinging South. Now put the perpetual motion holder under the North Pole hanging magnet, raise up slowly, then you will see the North Pole hanging magnet swinging North. This experiment shows without any doubt that the North and South Pole individual magnets are running in the same direction as those in the copper wire, which came out of the car battery, and in both instances while the magnets are running ahead in whirling motion they used the right-hand twist.

Get that Alnico magnet, and make it so you can turn it around if possible more than two thousand revolutions a minute. Connect the light bulb with the perpetual motion holder, put it on the spinning Alnico magnet in the hole between prongs and the square iron bar, now spin the Alnico magnet around and see how much of the light you get. Now take the iron bar off, then you will get more of the light. It shows that if it is closed, some of the magnets which are in the iron prongs will run around in an orbit, and will not come out, but when the orbit is broken then they will run in the coil, and the result will be more light.

Put a paper box with plenty of iron filings in it on the horizontally spinning Alnico magnet, then you will see how the spinning magnet builds up ridges and ditches. Now put the magnet so that it can be turned vertically. Spin the magnet, then you will see the filings running against the motion and building up ridges and ditches. Put on finer filings, then there will be finer ridges and ditches. Spin one way and then the other way, then you will have some rough idea how magnets build up the matter.

You made magnetic currents in three different ways, but in principle they all were made exactly in the same way. Magnetic currents are made by concentrating, then dividing and then shifting the existing North and South Pole individual magnets from one place to another. Now I will illustrate how my best machine is doing it. I will use only one coil, and one U shape permanent magnet without using the winding that the machine uses to increase the permanent magnet strength. If you had a permanent magnet that the coil you use in the electric magnet would go in between the prongs of it, then that would be good to demonstrate, but if you have not, then use the same one you have. Get an iron core the same dimensions as in the three-inch coil, but long enough to go between the permanent magnet prongs. Wind the same number of turns and connect with the light bulb. Fasten the U shape permanent magnet very good, bend up, prongs down, North Pole North, South Pole South. Now push the coil through the prongs from West to East. Do it fast, then there will be light in the bulb, now push the coil and stop in middle, and then push again, this time you will have two lights while the coil went through the magnet prongs only once. You had two lights the first time also, but you did not notice they came in quick succession. When you pushed the coil's middle up to field magnet's middle the currents ran in one direction, and when you pushed the coil away from the field magnet's middle, then the currents reversed, then ran in the other direction. That is why you got two light flashes while the coil passed through the field magnet only one time.

Here is the way in which the North and South Pole individual magnet currents ran while you pushed the coil from West to East through the field magnet. Take the core out of the coil, wind one layer of wire on the core and make it so that the North side of the winding wire's end points East and South side of the winding wire's end points West. When you pushed the coil to the middle of the field magnet, the North Pole magnet current came out of the wire end that is pointing East, and the South Pole magnet current came out of the wire end that is pointing West, but when you pushed the coil away from the middle of the field magnet the currents reversed, then North Pole magnet current came out of the coil's wire end that is pointing West and South Pole magnet current came out of the coil's wire end that is pointing East. With the same winding if the North Pole field magnet had been southside, and South pole field magnet northside, then the running of the currents would be reversed.

When currents reverse they reverse the magnet poles in the coil. Every time when the coil is approaching the field magnets, the currents which are made in the coil during that time are making magnet poles in the coil's core ends, the same as those field magnet poles they are approaching, but during the time the coil is receding those currents are making the coil's magnet poles opposite to the field magnets they are receding from. While you have the small coil handy I will tell more about magnets. Run South Pole magnet current in the wire end that points West, and North Pole magnet current in the wire end that points East. Now North end of the coil is South Pole and South end of the coil is North Pole. Now run North Pole magnet current in West end of the wire, and South Pole magnet in East end of the wire. This time the North end of the coil will be North Pole, and South end of the coil the South Pole.

You made the one-inch-long magnets with a single wire, but if you had the same size of wire in a coil you now have and would put a bigger steel bar in the coil then you would have a bigger and stronger magnet, but to make a stronger magnet yet, you would have to wind more layers on top of the coil that you have now. When you were making the small magnets with a single copper wire you wasted too many North and South Pole individual magnets. You only got in the steel wire very small part of the magnets that came out of the copper wire. You are still wasting the North and South Pole magnets. You do not get one-half of the magnets in the steel or iron bar from those which are in the coil.

To get more magnet out of a coil put the coil in steel or iron tube, then the tube outside the coil will be a magnet the same as the coil's core, but the magnet poles will be opposite, it means at the same coil end if the core end is North Pole the tube end will be South Pole. In this way you will get almost again as much magnet out of the coil and in the core and tube. You can do better yet, join one end of the coil's core end with the same metal, joining core with tube, make two holes in end of metal for the coil wire ends to go out, fasten a ring on top, now you

have the most efficient electric magnet for lifting purposes. It wastes
no magnets that come from your battery or dynamo.

Take the coil out of the electric magnet, run the currents in the
coil, put a hard steel bar one end to the coil's North Pole, hold awhile,
take away, now the bar is a permanent magnet. That end at coil's side
is South Pole magnet, and the other North Pole magnet. Now this permanent magnet can make other hard steel bars in permanent magnets
but every magnet that it makes will be a weaker magnet than itself. The
coil made this permanent magnet in the same way that the permanent
magnets are making other permanent magnets. Put this permanent
magnet in the coil's hole, reverse it, put bar's North Pole end in coil's
South Pole end, run current in the coil for a while, take the bar out, now
you have a stronger permanent magnet, but the poles are reversed. This
shows that the stronger magnet can change the weaker magnet.

When you were pushing the coil through the U shape magnet you
got two flashes in the light bulb with one passage through the U shape
magnet, and I showed you from which ends of coil's wire the currents
came out while they made the flashes. Now I will make so you can ac-
tually see that it is in the way I told you. Take the light bulb off the
coil, put the core in it, connect the coil with a loop that would reach
six feet East from the U shape magnet. Keep the loop end a foot apart,
stretch South side wire straight, make it so it cannot move. Get those
little hanging magnets which hang one end up, the other down, hang
the South Pole magnet on the loop wire, now push the coil through
the U shape magnet and watch the hanging magnet. First it will swing
South, then North. Now hang North Pole magnet on the wire, watch
again while you are pushing the coil through the U shape magnet, this
time first it will swing North, then South. Hang both magnets, watch
again and you will see that both magnets at the same time first they
swing to their own side and then to the other side. If the hanging magnets do not swing while you are pushing the coil through the U shape
magnet, then the U shape magnet is not strong enough. The U shape
magnet should be strong enough to lift twenty pounds. You can put two
magnets together or use electric magnet, and still better you can put
the coil in electric magnet, then you won't have to push it. Then you
can sit down and tap the battery and see the hanging magnets swinging.
All currents are made in the same way by filling the coil and iron core
with North and South Pole individual magnets and then giving enough
time for the magnets to get out and then start over again. If you want
to use the electric magnet be sure that the North Pole is in North side,
and the South Pole in South side, and put the coil in the prongs in the
same way as it is now.

Now I will tell you what happened to the U shape magnet while you pushed the coil through it from West to East. Set up the three-foot magnet so it can turn, put the coil with core in it in the U shape magnet, now approach the three-foot magnet's South Pole with the U shape magnet's South Pole. As soon as the three-foot magnet begins to move you stop and mark the distance. Take the coil away, approach again as soon as the three-foot magnet begins to move away, then stop and mark the distance, then you will see how much strength the U shape magnet lost while you were pushing the coil in and halfway out of the U shape magnet. The U shape magnet was losing its strength up to the time it began to break away from the iron core, but during the time the U shape magnet broke away it regained its strength. The breaking away from the iron core recharged the U shape magnet, then it became normal again and ready for the next start. During the recharging the new supply of magnets came from the air or the earth's magnetic field.

Now we see how the magnetic currents are made by the U shape magnet. You already know that before the coil got in between the U shape magnet prongs those little individual magnets were running out of the U shape magnet prongs in all directions, but as soon as the coil's core came in effective distance from the U shape magnet's prongs then these little individual magnets began to run in the core and coil and kept running until the core broke away from the U shape magnet prongs. Now you see those little individual magnets ran out of the U shape magnet and ran in the soft iron core, but the soft iron core never held the magnets, it pushed them out. To prove it you put five or six thin iron strips on edge, slant just so they will not flop over, now approach to the ends of those strips with a magnet and you will see they flop over, hold the strips a little loose by the ends, then they will spread out. I think this is enough to show that the soft iron never held those magnets. It pushed them out. As soon as those little individual magnets get pushed out of the soft iron core then they run in the coil. When they run in the coil they are in bulk form. The coil's part is to divide those little individual magnets from bulk form in small paths. The coil is not necessary to make magnetic currents. Currents can be made with a single wire. The coil is necessary to increase the amount and strength of the currents. The coil is similar to any cell battery. One cell alone does not amount to anything. To be good, many cells have to be in a battery. The same in a coil to be good many turns have to be in a coil.

When the magnets that are in bulk form enter the coil then the coil divides them in small paths. It is done in this way. When the bulk magnets enter the coil they fill the coil's wire with North and South Pole individual magnets. North Pole magnets pointing toward South Pole

U shape magnet and South Pole pointing toward North Pole U shape magnet. Now the wire in the coil is one continuous magnet. One side of the wire is South Pole and the other North Pole. Now we have those little North and South Pole individual magnets in the wire, but they are not running in the way we want. They are running across the wire. We want the magnets to run through the wire lengthwise, but there is only one way to do it, we have to increase the number of those North and South Pole individual magnets. To do it the coil will have to approach and enter the U shape magnet, but when the coil reaches the middle of the U shape magnet the limit is there so the running of the currents stop. In the core and the coil there is plenty of those little magnets, but they stopped to run through the wire lengthwise, now they run only across the coil's wire, to make the magnets run in the wire lengthwise again the coil will have to get away from the U shape magnet. As soon as the coil begins to move away from the U shape magnet, then those little North and South Pole individual magnets begin to run again through the wire lengthwise, but in opposite direction until the magnets in the iron core are gone.

I told you that the coil is a magnet during the time the currents are made, now I will show you. Get a small paper box to go in between the prongs of the U shape magnet, put iron filings in it. Wrap six-inch long soft iron wire with paper, put the wire in box in iron filings, now put the box between the U shape magnet prongs. Raise the wire up, then you will see filing strands clinging to the insulated iron wire. Raise the wire up slowly, then the filing strands will sag and fall, take the box out, put the wire in the filings again, raise up and you will see that the wire is no magnet, but during the time it was between the U shape magnet prongs it was a magnet. This shows that during the time the coil moves through the U shape magnet the coil becomes a magnet, but its function is double. Some individual North and South Pole magnets run through the coil's wire crosswise, and some run through the coil's wire lengthwise.

Maybe you think that it is not fair to use iron wire to demonstrate how magnetic currents are made, but I can tell you that if I do not use iron core in the coil I can make more of the magnetic currents with soft iron wire coil than I can with copper wire coil, so you see it is perfectly good to use iron wire to demonstrate how magnetic currents are made. You can do the same thing with the copper wire in using iron filings, but only on a smaller scale.

You saw how the magnets are running through a wire crosswise. Now I will tell you how they are running through the wire lengthwise. Before the magnets start to run through the wire lengthwise they are lined up in a square across the wire, one side of the wire is North Pole magnet side and the other side is South Pole magnet side. When the coil begins to approach the middle of the U shape magnet and the currents begin to run then the magnets which are in the wire begin to slant, North Pole magnets pointing East the same as the coil's wire end, where the North Pole magnet current came out and South Pole magnets pointing West the same as the coil's wire end where the South Pole magnet current came out. When the coil reaches the middle of the U shape magnet then the currents stop to run. Now the North and South Pole magnets are pointing across the wire again. When the coil begins to move away from the middle of the U shape magnet and the currents begin to run then the magnets which are in the wire begin to slant, but this time the North Pole magnets are pointing West the same as the coil's wire end where the North Pole magnet current came out and South Pole magnets pointing East the same as the coil's wire end where the South Pole magnet current came out. When the coil moves out of the U shape magnet's effective distance the currents running stop. This is the way the alternating currents are made.

When the individual North and South pole magnets are running through a wire lengthwise they are running in slant and whirling around while running ahead. You can see the slant by watching the sparks when you are putting together and pulling away soft iron wire ends which are connected to the battery by their other ends. To see how the currents are running out of the coil's wire watch those six one-inch long magnets which lie on the glass. Put those magnets together with ends even, then let them loose, then you will see that they will roll away and if the magnets be stronger then they will roll away farther. This is the way the North and South Pole individual magnets are running out of the coil's wire lengthwise. The reason the North and South Pole individual magnets do not run across through the coil's wire as fast out as they run in while the coil is between the U shape magnet, the coil's wire is insulated, there is an air space around every wire and as it is known that the dry air is the best obstruction for the magnets to go through and as you know the coil is well insulated so the damp air does not get in. It is well known that it is many times easier for the magnets to run in metal than in air, now you see when the magnets run in the wire they hesitate to run out of the wire across the same way as they came in, so more of the new magnets are coming in the wire crosswise, then they can get out crosswise, so they get pushed out through the wire lengthwise. Now you know how the alternating magnetic currents are made.

You have been wondering why alternating currents can run so far away from their generators. One reason is between every time the currents start and stop there is no pressure in the wire so the magnets from the air run in the wire and when the run starts there already are magnets in the wire which do not have to come from the generator, so the power line itself is a small generator which assists the big generator to furnish the magnets for the currents to run with. I have a generator that generates currents on a small scale from the air without using any magnets around it.

Another thing, you have been wondering how a U shape permanent magnet can keep its normal strength indefinitely. You know the soft iron does not hold magnets, but you already have one that holds it. It is the perpetual motion holder. It illustrates the principle how permanent magnets are made. All that has to be done is to start the magnets to run in an orbit, then they will never stop. Hard steel U shape magnets have a broken orbit, but under proper conditions it is permanent. I think the structure of the metal is the answer. I have two U shape magnets. They look alike, but one is a little harder than the other. The harder one can lift three pounds more than the softer one. I have been tempering the other steel magnets, and have noticed that the harder the steel gets the smaller it becomes. That shows that the metal is more packed and has less holes in it so the magnets cannot pass through it in full speed, so they dam up in the prong ends. They come in faster than they can get out. I think the ability for the soft steel welding rod to hold magnets is in the metal's fine structure.

The reason I call the results of North and South Pole magnet's functions magnetic currents and not electric currents or electricity is the electricity is connected too much with those non-existing electrons. If it had been called magneticity then I would accept it. Magneticity would indicate that it has a magnetic base and so it would be all right.

As I said in the beginning, the North and South Pole magnets they are the cosmic force. They hold together this earth and everything on it, and they hold together the moon, too. The moon's North end holds South Pole magnets the same as the earth's North end. The moon's South end holds North Pole magnets the same as the earth's South end. Those people who have been wondering why the moon does not come down all they have to do is to give the moon one-half of a turn so that the North end would be in South side, and South end in the North side, and then the moon would come down. At present the earth and the moon have like magnet poles in the same sides so their own magnet poles keep themselves apart, but when the poles are reversed, then they will pull together. Here is a good tip to the rocket people. Make the rocket's head strong North Pole magnet, and the tail end strong South Pole magnet, and then shut to on the moon's North end, then you will have better success.

North and South Pole magnets are not only holding together the earth and moon, but they are turning the earth around on its axis. Those magnets which are coming down from the sun they are hitting their own kind of magnets which are circulating around the earth and they hit more on the East side than on the West side, and that is what makes the earth turn around. North and South Pole magnets make the lightning, in earth's North hemisphere the South Pole magnets are going up and the North pole magnets are coming down in the same flash. In the earth's South hemisphere the North Pole magnets are going up and the South Pole magnets are coming down in the same flash. The North lights are caused by the North and South Pole magnets passing in concentrated streams, but the streams are not as much concentrated as they are in the lightning. The radio waves are made by the North and South Pole magnets. Now about the magnet size. You know sunlight can go through glass, paper and leaves, but it cannot go through wood, rock and iron, but the magnets can go through everything. This shows that each magnet is smaller than each particle of light.

A Book in
Every Home

Containing Three Subjects:

Ed's Sweet Sixteen,
Domestic & Political Views

by

Edward Leedskalnin

PREFACE

THE Coral Castle of Florida is one of South Florida's historical and
famous landmarks. The story of this beautiful and fascinating wonderland of coral rock reads like a fairy tale, and began over fifty years
ago when an obscure Latvian immigrant named Edward Leedskalnin
came to South Florida. He chose this sparsely settled section of Florida
because he "wanted to get away from the world". Here he began many
years of self-induced hard labor in order to forget his unrequited love
for his one and only sweetheart. This young girl, whom he always re-
ferred to as his "Sweet Sixteen", jilted him for another on the eve of
their wedding day in his native Latvia. An extremely sensitive soul, he
was deeply hurt and made the decision to leave his native country and
seek solace elsewhere. Emigrating to America in 1921, he went directly
to California, staying there but a few months, and then moving on to
Texas. From Texas he made his way to Florida. He did not stop until
he came within sight of the very southernmost tip of Florida, south of
Florida City.

Here on the edge of the Everglades he apparently felt he was far
enough away from people, and settled down on a small plot of land which
a generous neighbor permitted him to use. He built a house from logs
cut from the surrounding pine trees, and coral rock which he quarried
on the premises. Thus began the strange and unusual project he set for
himself, interrupted only by his moving to a larger plot of ground three
miles north of Homestead, which after twenty-five years of continuous
labor culminated in the completion of what can only be described as a
"Coral Castle", breathtaking in scope and imagination, unbelievable for
skill and patience required.

Today, the Coral Castle of Florida, as it is known, lies athwart U. S.
Highway No.1 about twenty-five miles south of Miami. The Coral Castle

25

is no ordinary structure. It is set on a ten-acre tract of land, the castle proper being surrounded by an eight-foot high wall made of huge blocks of coral rock, each weighing several tons. The tower contains 160 tons of coral rock, each block making up this huge building weighing nine tons, the first floor of which Leedskalnin used as a workshop and the second floor housed his living quarters. An air of mystery prevailed about these quarters since no one was permitted entry. Behind the huge walls, in beautiful settings, are the most marvelous and fantastic pieces of coral rock furniture, and movable objects which he created from his fertile imagination. There are rocking chairs weighing thousands of pounds so delicately balanced they move at the touch of a finger. Couches, beds, chairs, tables of all sizes and shapes, including one table hewn from solid coral rock into the shape of the State of Florida, and another cut into the shape of a heart with a beautiful ever blooming floral centerpiece growing out of the center of this rock table. There are huge crescents atop walls twenty feet high, an obelisk reaching up to the sky weighing twenty-eight tons that Leedskalnin set in place by the use of simple hand tools. There is an ingenious Polaris telescope carved out of the coral rock rising twenty-five feet into the air. With this telescope he would study astronomy nightly under the Florida skies. As in most castles, there is a subterranean well, with a circular staircase carved out of the rock leading down into the water.

Upon entering the Coral Castle it is necessary to go through the swinging gate, a triangular block of stone weighing three tons that turns, permitting you to pass in. In the center rear wall is a huge stone gate weighing nine tons reminiscent of ancient castles that is so perfectly pivoted and balanced that a child can turn it. On the opposite side is constructed an iron ladder which allows you to get up onto the huge Crescent of the East, from which one can overlook the entire park.

Leedskalnin never forgot his "Sweet Sixteen", and much of the architecture he created was designed with the thought uppermost in his mind that some day she would return to him. This is apparent from the complete bedroom hewn out of the coral rock furnished with twin beds, a pair of children's beds, as well as a baby cradle; also from the beautiful patio for children to play in, called the "Grotto of the Three Bears", containing all the furniture in the famous fairy tale, including a huge coral porridge bowl. He never married, living the life of a true hermit in this coral wonderland of his.

The Coral Castle has become a familiar sight over the years to travelers on U. S. Highway No. 1, and many, intrigued by this huge coral edifice, have taken the time to explore it, while others raced by wondering at its incongruity in this modern world of ours. Those who stopped were met by the man who constructed this monumental work, and guided around while he explained its many mysteries. But one thing he never told anyone was how he ever was able to move the huge coral rocks weighing up to thirty-five tons which he excavated single-handed. When asked, he replied simply that he knew the secrets used in the building of the Pyramids of Egypt. Whatever the secrets or principles of construction he used, they died with him when he passed away in a Miami hospital in December 1951. It can be said his knowledge of the practical use of pulleys, and leverage was not exceeded by the ancient Greeks or Egyptians.

Edward Leedskalnin's remarkable versatility is shown also by his studies and experiments in magnetic current, and he evolved and published many theories on the subject which he claimed are more accurate than any of the known thoughts on magnetism. For instance, his theory of Cosmic Force is remarkably akin to Einstein's latest theory of Unity. In the pamphlet he published in 1946, he states, "the North and South poles are the cosmic force. They hold together the earth and everything on it—turning the earth around on its axis".

In an article reprinted in the Miami Daily News, he writes, "Every form of existence, whether it be rock, plant or animal life, has a beginning and an end, but the three things that all matter is constructed from has no beginning and no end. They are the North and South poles' individual magnets, and the neutral particles of matter. These three thing are the construction blocks of everything". In addition to all these studies, he found time to write and publish this booklet called, "A Book in Every Home", giving his opinions on life in general based on his observations of people. Here one can gain an insight to Leedskalnin's ideas as he states, "An educated person is one whose senses are refined. We are born as brutes, we remain and die as the same if we do not become polished. But all senses do not take polish. Some are too coarse to take it. The main base of education is one's 'self-respect'. Anyone lacking self-respect cannot be educated. The main bases of self-respect is the willingness to learn, to do only the things that are good and right, to believe only in the things that can be proved, to possess appreciation and self-control".

The Coral Castle is known to countless thousands, and is considered one of the wonders of the world, baffling engineers and laymen alike. A masterpiece of rugged and permanent beauty, it is a landmark of the Redlands district of South Florida, and Edward Leedskalnin himself has become a legendary figure. Compared by many authorities to the Stonehenge in England, and the Pyramids of Egypt, Coral Castle of Florida is now the mecca for visitors from all over the world. Upon seeing the Coral Castle, a prominent minister stated, "After seeing this, I can give a great sermon on the quality of patience". A noted visitor, Dr. David Fairchild, founder of Fairchild Tropical Gardens, wrote in the guest book, "'This is one of the most interesting spots in Florida. I am glad I knew Edward Leedskalnin".

Thus, although he never again saw the girl who inspired him, the Coral Castle of Florida stands now and forever as a monument for all to see and marvel at the genius, imagination, and skill of this strange but brilliant man, Edward Leedskalnin.

Author's Preface

Reader, if for any reason you do not like the things I say in this little book, I left just as much space as I used, so you can write your own opinion opposite it and see if you can do better.

The Author

ED'S SWEET SIXTEEN

To THOSE MORE THAN fifteen thousand people who have seen Ed's Place, I told about Ed's Sweet Sixteen. Now, I will tell you why I did not get the girl.

In Ed's Place, there was a lasting fame for a girl's name but it would have taken money to put the fame upon her. The trouble was that I did not have the money and did not make enough. That was the reason I could not look for a girl.

Now, I am going to tell you what I mean when I say "Ed's Sweet Sixteen". I don't mean a sixteen year old girl, I mean a brand new one. If it had meant a sixteen year old girl, it would have meant at the same time, that I made money for the sweet sixteen while she was making love with a fresh boy.

I will furnish all the love making to my girl. She will never have to seek any from anybody else, for I believe that there is not a boy or a man in this world good enough to be around my girl and I believe that the other men also ought to have enough self-respect so that they would think that I am not good enough to be around theirs.

Anything that we do leaves its effect, but it leaves more effect upon a girl than it does upon a boy or a man, because the girl's body, mind and all her constitution is more tender and so it leaves more impressions— and why should one want to be around anybody's else impressions?

A girl is to a fellow the best thing in this world, but to have the best one second hand, it is humiliating.

All girls below sixteen should be brand new. If a girl below sixteen cannot be called a brand new any more, it is not the girl's fault; the mamma is to blame! It is the mamma's duty to supervise the girl to keep those fresh boys away.

In case a girl's mamma thinks that there is a boy somewhere who needs experience, then she, herself, could pose as an experimental station for that fresh boy to practice on and so save the girl. Nothing can hurt her any more. She has already gone through all the experience that can be gone through and so in her case, it would be all right.

But all the blame does not rest on the mamma alone. The schools and the churches are cheapening the girls! They are arranging picnics—are coupling up the girls with the fresh boys—and then they send them out to the woods, parks, beaches and other places so that they can practice in first degree love making.

Now, I will tell you what the first degree love making is. The first degree love making is when the fresh boy begins to soil the girl by patting, rubbing and squeezing her. They start it in that way but soon it begins to get dull and there is no kick in it, so they have to start in on the second degree and keep on and then by and by, when the right man comes along and when he touches the girl, then he touches her like dead flesh. There is no more response in it because all the response has been worked out with those fresh boys. Why should it be that way?

Everything we do should be for some good purpose but as everybody knows there is nothing good that can come to a girl from a fresh boy. When a girl is sixteen or seventeen years old, she is as good as she ever will be, but when a boy is sixteen years old, he is then fresher than in all his stages of development. He is then not big enough to work but he is too big to be kept in a nursery and then to allow such a fresh thing to soil a girl—it could not work on my girl. Now I will tell you about soiling. Anything that is done, if it is with the right party it is all right, but when it is with the wrong party, it is soiling, and concerning those fresh boys with the girls, it is wrong every time.

Now, how can you find out if I am right? Pick out any girl you want but do it before she has anything to do with anybody—as soon as she begins to couple herself with somebody. You watch her every day and some day you will see the girl coming home with a red face. One's face is a window for other people to look in on and when it turns red it shows that there was something done that her moral conscience told her should not have been done.

It is shocking to imagine that someone else produced that red face to my girl. In such a case she could not be one hundred per cent sweet. As soon as a girl acquires experience the sweetness begins to leave her right away. The first experience in everything is the most impressive. It should be reserved for the permanent partner—the less of the new experience is left, the cheaper the affair will be.

That is the reason why I want a girl the way Mother nature puts her out. This means before anybody has had any chance to be around her and before she begins to misrepresent herself. I want to pick out the girl while she is guided by the instinct alone.

When I started out in doing things that would make it possible for me to get a girl, I set a rule in my behaviour to follow:

The sweet sixteen had to be a beginner and a likeable girl and with a mild disposition; I had to be deserving of her. Everybody's sweet sixteen should be so high in one's estimation that no temptation could induce one to act behind her back. I always have wanted a girl but I never had one.

The reason why is that I knew it would produce several conditions and leave their effects, but I did not want any effects from past experience left on me and my sweet sixteen.

A girl will economize, go hungry and endure other hardships before she will put on another girl's dress to wear. I will put gunny sacks on before I will wear another man's clothes, and this is only a step from having another fellow's girl or another girl's fellow.

Having such a case the present possessor would have to clean up the past performer's effects. Now you see, to clean up the other person's leavings, it is humiliating, so it would be a cheap and undesirable affair. I want one hundred per cent good or none. That is why I was so successful in resisting the natural urge for love making.

Now about sweet and how sweet, a girl can be one hundred per cent sweet to one only and no more. To illustrate, suppose we are two men and a girl together somewhere and some one else would ask if she is sweet and we both would say she is sweet. But let her act very friendly with the other fellow and then if someone should ask if she is sweet, I would say that she is not. Now you see her friendly action with the other fellow produced a change in me and it would produce a change in any other normal man.

We always strive for perfection. We are only one-half of a perfect whole, man is the bigger and stronger half and the woman is the other. To be perfect there must be two, but where there is two there is no room for more, so the third party is left to go somewhere else with sour feelings.

A normal male is always ready to strive for perfection, the female is not. It is not only with human beings, it is the same with every living thing. If you watch a flock of chickens, where there is a rooster, and if you add another rooster, you will see them fight to death. One will have to go or be killed and this is the same thing with the other living things.

Lower forms of life are guided by instinct alone so the present only comes into consideration. As soon as the other male is chased away, the female is as good as she ever was, but with us it is different. We are guided more by reason and thought than by instinct and so the present, past and future come into consideration. Now, if it is not good today, it was not good yesterday and it won't be good tomorrow. That is why an experienced girl cannot be one hundred per cent sweet.

According to my observation the girls are wrong in looking for their permanent partners. They are too quick. By being too quick, they only get those fellows with quick emotions. All quick emotions are irresponsible and short lasting.

There are two kinds of love—sensual and sentimental. Sensual love has the present and little future only. The sentimental love has the present, past and future, so it is more desirable. It will be slower but will last longer. Now, girls, when any fellow jumps quickly at you, you had better keep away from him. He is acting wholly selfish. He has no consideration that the action would do any good to you. You are the weakest side, so you should have the better deal and if you don't get the better deal, there is a little brute in him and it may come very hard to train it out of him. The fellow who makes an advance toward you, and if he won't state what the eventual purpose will be, he is not a gentleman. All men should know that the girl's primary purpose is to find a permanent partner while they are young. Those fellows who fail to see this are not desirable to have around.

Girls below sixteen should not be allowed to associate with the boys, they are practicing in love making, such a thing should be discouraged. Love making should be reserved for their permanent partners. With every love making affair, their hearts get bruised and by the time they grow up, their hearts are so badly bruised that they are no more good.

Boys and girls start out as friends and finish as disappointed lovers, now let me tell you. Male and female are never friends, a friend will not want anything from a friend, but a boy or a girl, one or the other, sooner or later, will ask for a little kiss, so they are not friends, they are lovers.

Let's see what happens when they are selecting their partners while they are young. They select their partners on account of good looks. The liking for the good looks remains but the good looks change and they change so much in ten years that you would not recognize them if you had not seen them now and then—and the boy gets the best deal almost every time. By that time they are grown up. The girls will be faded so much that the fellow would not want her any more so then, any girl who associates with a fellow only five years older is headed for a bad disappointment. This all could be avoided with the right kind of an education.

Now, a few words about education. You know we receive an education in the schools from books. All those books that people became educated from twenty-five years ago, are wrong now, and those that are good now, will be wrong again twenty-five years from now. So if they are wrong then, they are also wrong now, and the one who is educated from the wrong books is not educated, he is misled. All books that are written are wrong, the one who is not educated cannot write a book and the one who is educated, is really not educated but he is misled and the one who is misled cannot write a book which is correct.

The misleading began when our far distant ancestors began to teach their descendants. You know they knew nothing but they passed their knowledge of nothing to the coming generations and it went so innocently that nobody noticed it. That is why we are not educated.

Now I will tell you what education is according to my reasoning. An educated person is one whose senses are refined. We are born as brutes, we remain and die as the same if we do not become polished. But all senses do not take polish. Some are too coarse to take it. The main base of education is one's "self-respect". Any one lacking self-respect cannot be educated. The main bases of self-respect is the willingness to learn, to do only the things that are good and right, to believe only in the things that can be proved, to possess appreciation and self control.

Now, if you lack willingness to learn, you will remain as a brute and if you do things that are not good and right, you will be a low person, and if you believe in things that cannot be proved, any feeble minded person can lead you, and if you lack appreciation, it takes away the incentive for good doing and if you lack self control you will never know the limit.

So all those lacking these characteristics in their makeup are not educated.

DOMESTIC

THE FOUNDATION OF OUR physical and mental behaviour is laid while we are in infancy, so the responsibility of our shortcomings rest upon our mothers and fathers, but mainly upon our mothers.

Today, I myself would be better than I am if my mother and father had known how to raise me and the same is true for almost everybody else.

At the first contraction in any part of your body, you will never notice any mark, but keep it up and some day you will see a crease, and it will be permanent. We all want to look and act the best that we know how, but we cannot learn from ourselves so we have to learn from others.

In my thirty years of studying conditions and their effects I have come to the conclusion that I can tell pointers to the people that would be a good help to them. That is why I wrote this little book.

To accomplish good results, the mothers will have to keep good watch on their darlings until they acquire the natural ambition to shine, and the girls should be more carefully watched than the boys, because the girl's looks are her best asset and should be cultivated.

Don't raise the girls too big by over feeding them and too curved by neglect.

People who want to shine will always have to restrain themselves, because if they don't, their actions won't be graceful. Even when one's looks are good, if he abandons restraint, the performance won't be good. It is more likely that the person himself won't notice but others will.

The first thing I notice about other people is, if there is something wrong and if it could be improved and the same must be true about other people noticing my defects and neglects. To correct those neglects, somebody will have to point them out, but to do it directly will not do, because they would think you are mean. That is why I want to point out the defects and neglects in this book.

The most striking neglect that comes to my attention is when one is smiling. A smile is always pleasing if it is regulated but without restraint, it is not. When smiling, the teeth only should be shown. As soon as you show the gums, it spoils the good effect. When showing the gums you are doing triple harm. First, the gums never look good; second, you are making too big creases in the side of your mouth and third, your lips come too wide apart. Especially should a girl be careful not to show too abnormally big mouth. Girls should do nothing that would impair their best looks. I have seen moving picture stars, public singers and others with their mouths open so wide that you would think the person lacks refinement, but if they knew how bad it looks they would train it out. No doubt they have practiced before a looking glass, but a looking glass does not show such an enormous opening, because while they are looking at the looking glass they are under restraint and so they really don't know how it looks while they are not watching themselves. In a looking glass you will never notice all your neglects and defects. They have to be pointed out by somebody else.

It is painful to hear other people pointing out our neglects and defects so do not entrust your friends to do it. Your friends may not always be your friends. The best way is to leave that to your own family. Your mother and father will do nothing to embarrass you. Your mother will do it better and it should be started while one is still a baby.

The first thing the mothers should do is to watch the baby's mouth so it is not hanging open. The mouth, by hanging open, stretches the upper lip and when kept open while growing, then when fully grown, the lips will not fit together any more.

Mothers should keep close watch on their children's behaviour. As soon as they notice some action and contraction that is not graceful, they should correct it immediately, because their actions leave their effects. To small children, it doesn't matter how ugly they look, but when they are grown up, the good looks will be the best thing, and one with a disfigured face cannot be satisfied with oneself. The foundation for one's best looks will have to be laid while one is small.

A graceful smile is pleasing but if it is not perfect, its pleasing effect is marred. To obtain better results, don't make the smile too big by opening the mouth too wide, drawing the lips over the gums, or drawing one side of the lip more than the other, or drawing both lips to one side and have them twisted.

Children should not be encouraged to smile too much, smiling in due time will produce creases in the sides of their mouths. It would be better to save the smiles till they are grown up. Children while they are growing should be watched, closely. They are stretching their mouths with their fingers and are jamming too big objects in their mouths and making too ugly faces. All those actions should be forbidden for their future's sake.

Eyes should be trained to look in the middle between both lids, never through the forehead. If this is done, it will produce creases in the forehead. When the lids of one eye are more narrow than the other it should be trained out and equalized.

In case one leg is shorter and one shoulder lower, they can be disguised so that other people would not notice it. In walking the toes should be carried a little out, by carrying the toes out one can walk better. Shorter steps would make the walking more graceful and those who stoop over, higher heels would help to keep the body more erect.

Girls should take smaller steps than boys. By taking smaller steps the body would not jump as much up and down or swing from side to side.

Mothers should study the other people's children and then pick out the best model from which to train their own child.

Everybody should be trained not to go out anywhere before somebody else has examined them to see if everything is all right. It would save many people from unexpected embarrassment.

POLITICAL

BEFORE I SAY ANYTHING about the government, let's establish a base for reasoning. All our ideas should produce good and lasting results and then anything that is good now would have been good in the past and it will be good in the future and it will be good under any circumstances, so any idea that does not cover all this broad base is no good.

To be right, one's thought will have to be based on natural facts, for really, Mother Nature only can tell what is right and what is wrong and the way that things should be.

My definition of right is that right is anything in nature that exists without artificial modification and all the others are wrong.

Now suppose you would say it is wrong. In that case, I would say you are wrong yourself because you came into this world through natural circumstances that you had nothing to do with and so as long as such a thing exists as yourself, I am right and you are wrong. Only those are right whose thoughts are based on natural facts and inclinations.

It is natural tendency for all living things to take it easy. You watch any living thing you want to, and you will see that as soon as they fill up, they will lie down and take it easy.

The physical comfort, the ease, that is the only thing in this world that satisfies. It cannot be overdone and it is the real base of all our actions. We all cannot take things easy because there is too much competition from other people only those who possess good management will succeed by exploiting domestic animals, machinery, other people and natural resources.

Everything will have to be produced that is consumed and to those who have to produce the things themselves, they are consuming the easy days are not coming to them.

It has been told to you that the government is for the purpose of protecting "life" and property, but it really is to protect "property" and life. Nobody wants your life but everybody wants your property.

In International dealings, when an army conquers the land, they don't want the people, they want the physical property and so do the thieves and the bandits. They want your money and property and if you will submit peacefully, they won't harm you.

Now you see, nobody wants you, they want your property so really the property is the one that needs the protection and not you. You are the protector yourself.

Government to be lasting will have to be just. This means it will have to protect all the property alike and all the property will have to pay equal taxes, which means big property, big taxes, and small property, small taxes. Government cannot exist without taxes so only those who pay taxes should vote and vote according to the taxes they pay.

It is not sound to allow the weaklings to vote. Any one who is too weak to make his own living is not strong enough to vote, because their weak influence weakens the state and a degenerated state cannot exist very long, but every state should be sound and lasting.

By voting, the voters dictate the state's destiny for times to come and then to allow such a weak influence to guide the state, it is not wise and so you see one should vote according to how he is carrying the State's burden.

Another unwise thing about equal voting is that it gives the loafers and weaklings the power to take the property away from producers and stronger people, and another unjust thing about equal voting is that it gives the loafers and weaklings the power to demand an easy life from the producers and leaders.

Self respecting producers will not stand such an injustice for long. It is not the producers' fault when one is too weak to make his own living. The producer's life is just as sweet as the weaklings and loafers life is to them. All people are independent so you see everybody will have to take care of themselves and if they cannot, they should perish and the sooner they perish the better it will be.

To be lasting, the government should be built in the same way as the Supreme power of the land "the army." Governments have been rising and falling but the army always remains. You know there is no equality in army and so there can be no equality in the state if you are not equal producer you cannot be an equal consumer.

Fifty per cent of the people don't want to lead, they want to follow. They want somebody else to furnish the money for their living expenses and as long as such a condition exists, they are not equal with their leaders. That is the reason why everybody should be put in the right place according to their physical and mental ability.

There is only one way to share the National income. It is by sharing the production and if you are not producing equally you cannot share equally. Nobody is producing anything for others. They are producing only for themselves.

People are individuals. For instance, if you want an excitement you will have to test the thrill yourself, or if you have a pain you will have to bear it yourself, or if you want to eat you will have to eat for yourself. Nobody can eat for you and so it is that if you want the things to eat you will have to produce them yourself and if you are too weak, too lazy, lack machinery and good management to produce them, you should perish and that is all there is to it.

Statistics on the Coral Castle of Florida

THE CORAL CASTLE OF Florida is the finest example of massive stone construction in the United States. A study of the immense sizes and weights of coral rock excavated, moved and used in its construction, in comparison with those used in many of the famous works around the world, establishes it as an authentic wonder of the world. The fact that it was accomplished entirely by one man makes it all the more remarkable.

Altogether there are approximately 1,000 tons of coral rock used in the construction of the walls and tower alone, a stupendous achievement for one man, unequalled in all history. In addition over 100 tons of coral rock were used in the carvings of the artistic objects throughout the entire park.

South and West Walls

There are 65 sections of stone weighing a total of approximately 420 tons, or an average of 6½ tons each in these walls.

EAST WALL

THERE ARE A TOTAL of approximately 240 tons of coral rock in this wall which contains the following:

Crescent of the East.................................. 23	Tons
Planet Mars 18	Tons
Planet Saturn.. 18	Tons
Obelisk.. 28½	Tons

NORTH WALL

Contains 149½ tons of stone. This wall contains the heaviest stone in the park, weight 29 tons; also the Polaris telescope, 25 feet tall and weighing 28 tons.

THE TOWER

The Tower contains 243 tons of coral rock made up of huge blocks of stone weighing up to 9 tons each.

The average weight of the stones used in the construction of the Coral Castle of Florida is greater than those used in the building of the Great Pyramid of Giza, while several of the stones used are taller than those found in the Stonehenge in England, and in weight exceed greatly the stones used in many of the other famous stone works throughout the world.

www.ingramcontent.com/pod-product-compliance
Lightning Source LLC
Chambersburg PA
CBHW031910200326
41597CB00012B/583